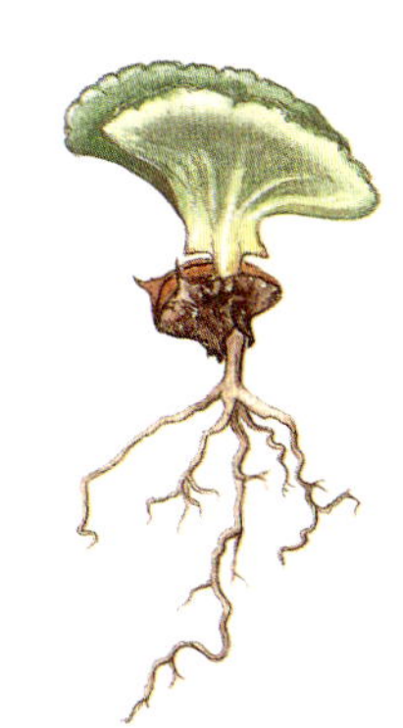

Banksia menziesii

Banksia menziesii

PHILIPPA NIKULINSKY

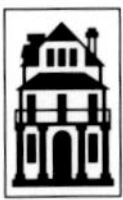

FREMANTLE ARTS CENTRE PRESS

First published 1992 by
FREMANTLE ARTS CENTRE PRESS
193 South Terrace (PO Box 320), South Fremantle, Western Australia 6162.

Editor B R Coffey. Designed by John Douglass. Production Manager Helen Idle.

Typeset in Goudy Oldstyle by City Typesetters, Western Australia.
Scanning by Prepress Services, Western Australia.
Printed by Dai Nippon, Japan.

National Library of Australia Cataloguing-in-publication data.

Nikulinksy, Philippa, 1942-
Banksia menziesii.
ISBN 1 86368 035 7.
1. Banksias in art. I. Title.
759.994

Fremantle Arts Centre Press receives financial assistance from the
Western Australian Department for the Arts.

THE ART OF THE BANKSIA

S/he to whom Nature
begins to reveal
her open secret
will feel an irresistable yearning
for her most worthy interpreter,
Art.

Johann Wolfgang von Goethe

One of the joys of being a botanical artist is the necessity for field trips into the bush; I don't need much persuasion to travel to be among the subjects of my work.

The *Banksia menziesii* (Firewood or Menzies' Banksia) has a wide distribution on the west coastal region of Australia, from the Murchison district – approximately five hundred kilometres north of Perth – to Pinjarra – seventy kilometres south of Perth. For this series of paintings I chose as my study specimens trees from a location north of Eneabba, a small town about two hundred kilometres north of Perth.

I travelled to this area many times over several months to look at trees during the various stages of their cycle, to sketch, and to collect specimens to take back to my studio for the detailed drawings and paintings.

As I became immersed in this work it felt as if I were becoming a part of the private and mysterious reality of the banksia. In trying to record this cycle – these frozen moments – and to express my response to it, I had the feeling of actually becoming part of the generative process. I am almost embarrassed to think back on how my excitement built with each succeeding image: the whole process was for me such a revelation. The absolute wonder of creation was overwhelming.

As I worked on the series I had the images lined up along the workbench in my studio. I would often just sit and look at them as they developed. I would think about these moments and wonder at the complex processes in the trees' life cycle, about the many transactions with soil, water, birds, animals, wind and rain. I would think, too, of how each moment represents the total of the transactions that preceded it.

In the months these paintings were on the studio wall many people saw them and, suprisingly, shared my excitement and sense of discovery. And because each person's perceptions are different – there is, I suppose, one reality but many interpretations – each found some individual understanding of the experience.

After seeing the paintings, an artist friend returned one day with an old zen saying she thought apt:

> *The realities of life are most truly*
> *seen in everyday things and actions.*

This reality can be seen in the two continuous cycles of the banksia – the sequence of flower to seed, earth, new plant and bud; whilst on each individual tree, a bud develops and then flowers as the bud for the next season begins. It is the way in which these processes may represent the experience of life, and give expression to a deeper understanding that I want to share.

Philippa Nikulinsky

Banksia menziesii

Dragon-pods swell,
feathered birth,
armoured transience.

Fay Zwicky

The pre-bud stage. These few nondescript,
wrinkled, hairy leaves and the untidy,
though delicate bracts forming at the end
of the stem may initially have little impact.
That is until we realise the hidden potential
that is stored in them. It is this potential,
this essence, that I have sought to capture.

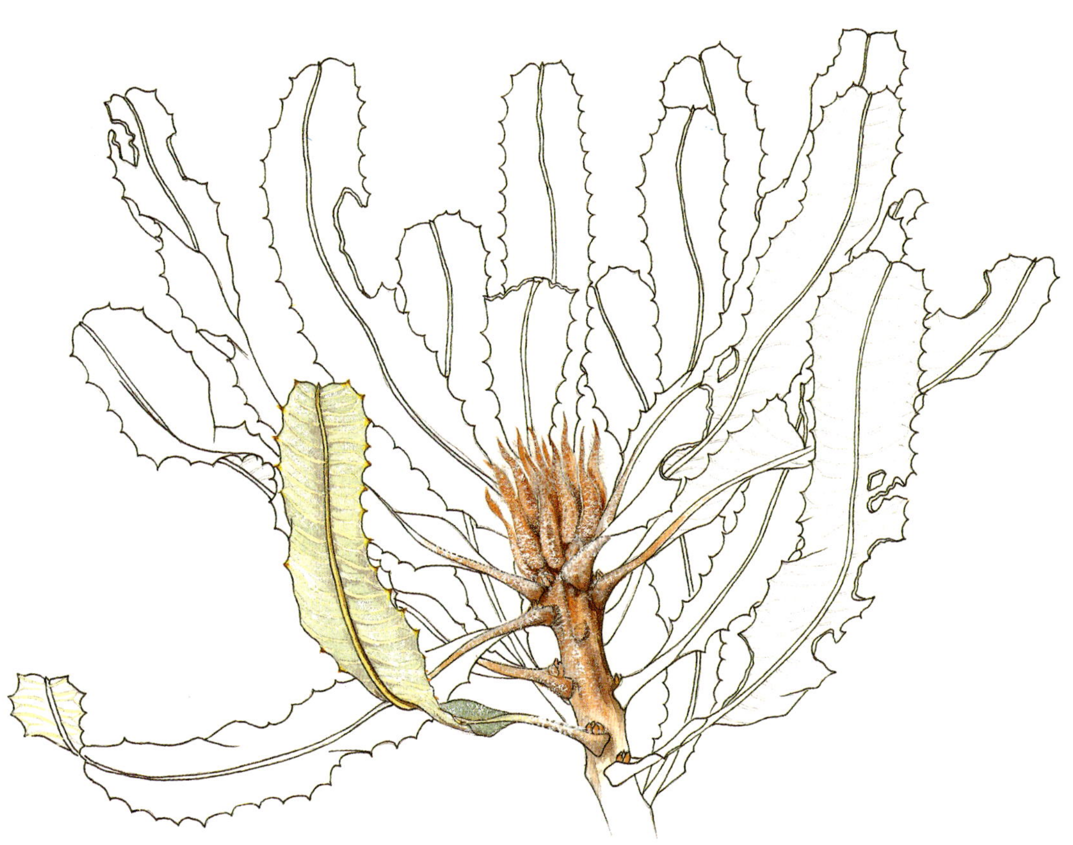

The perfection of the bud of the emerging
spike or inflorescence. I was attracted to
this stage because of the purity of the small
spike – soft, pale and symmetrical among
the rough, prickly surrounding leaves.

I often found myself distracted by the
leaves and would draw pages of them,
finding their curves, twists and
folds hypnotic.

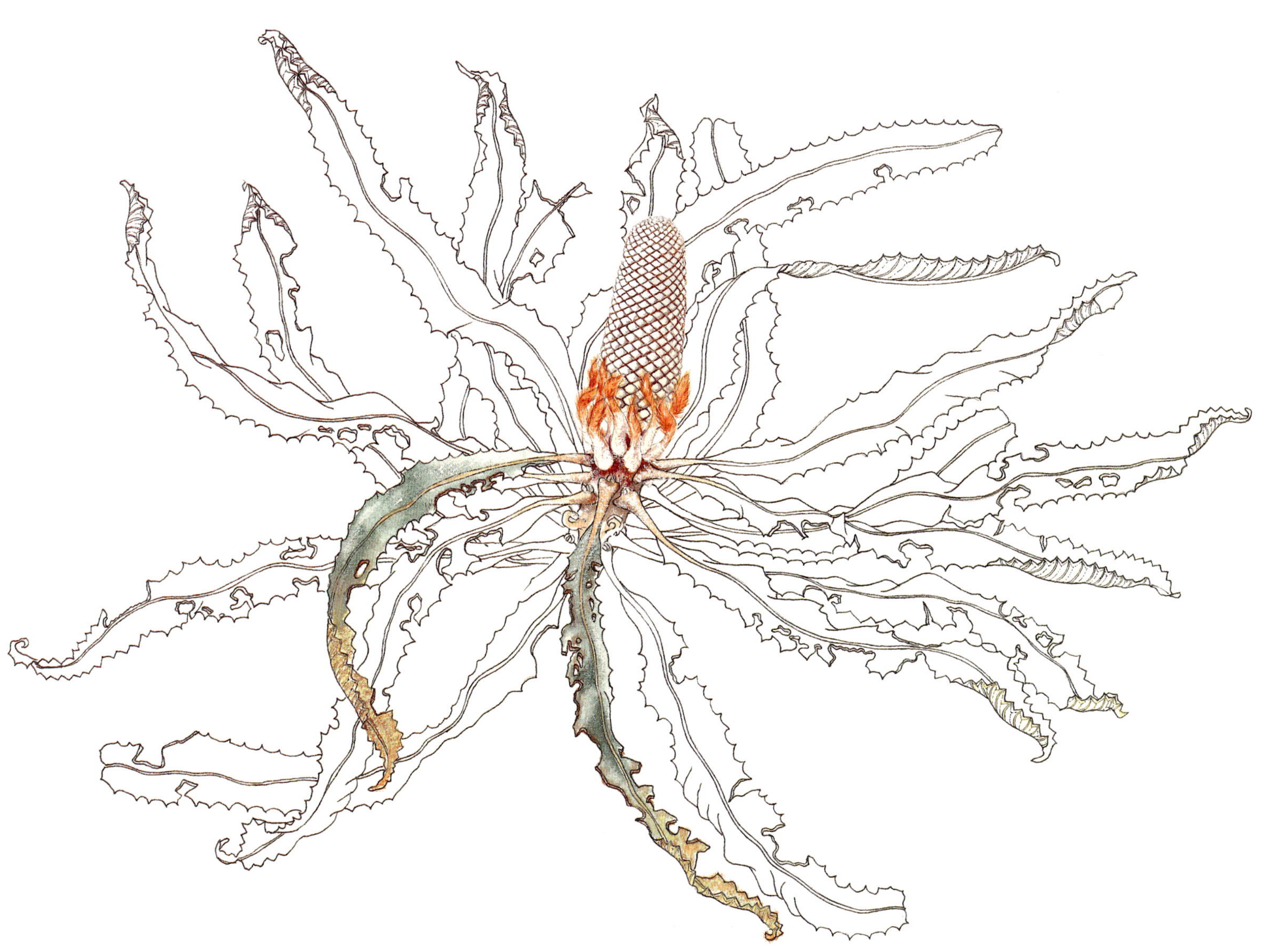

As the spike develops, the orderly pattern
of its surface slowly evolves from the
bottom, becoming more complex as
tiny flower buds begin to emerge.

When doing my drawings of the
developing spike I found it difficult to
concentrate on each segment of the pattern.
Each time I tried to focus on one part
my eyes would jump and follow
a different path.

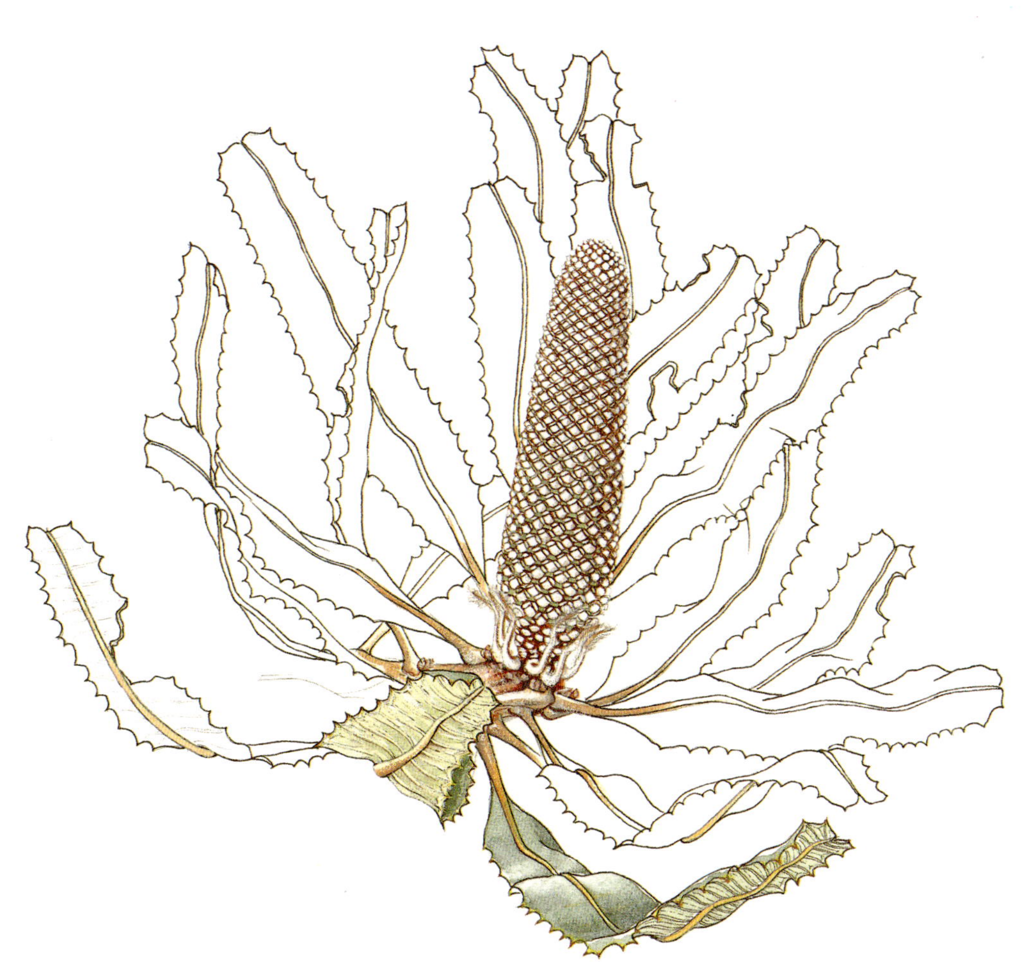

The emerging flower buds
gradually spiral upward around the
spike until it is completely covered.

In these early stages, the absolute order, the
tight patterns and the subtleties of colour
are what I found most fascinating.

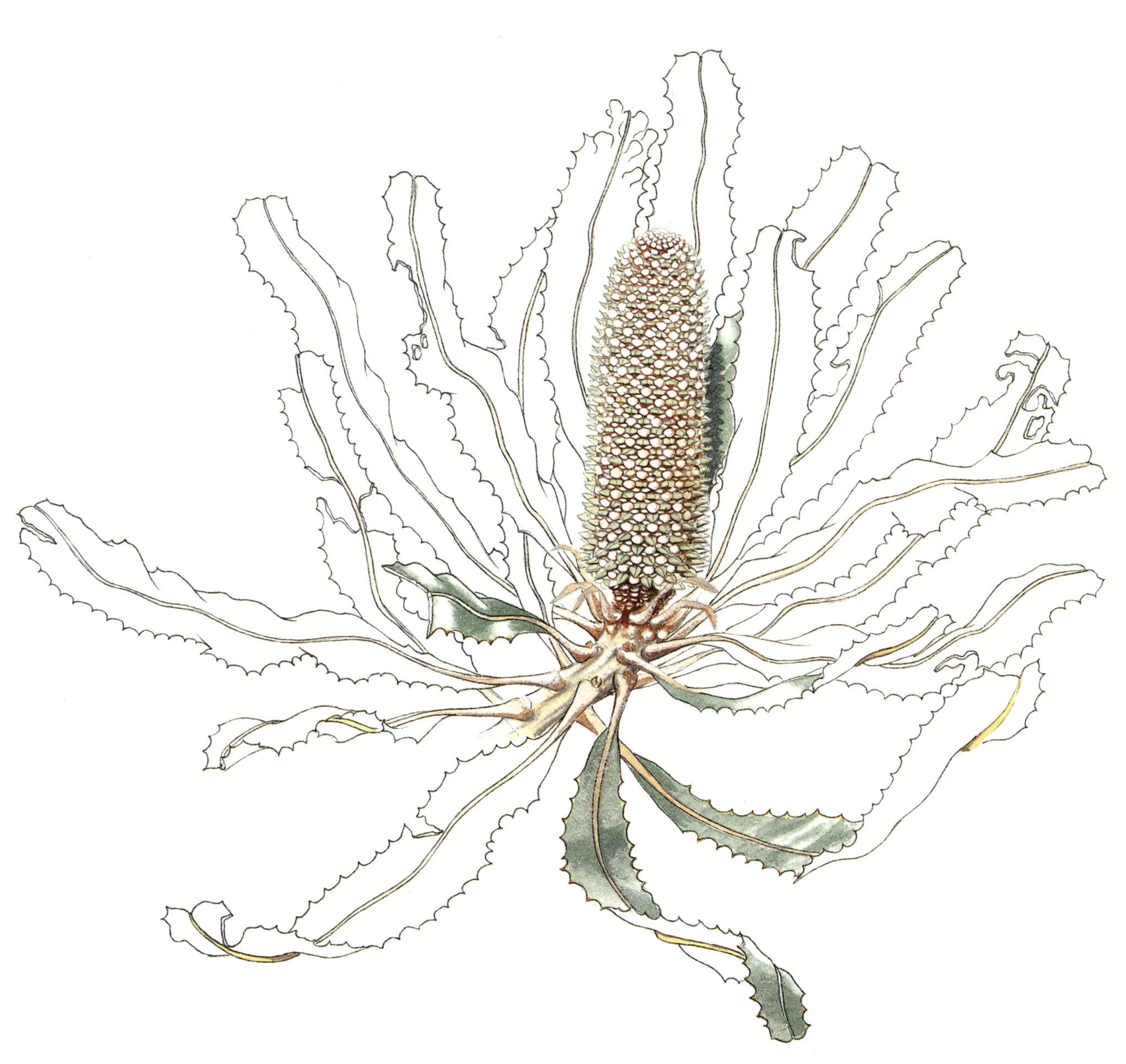

As the flower buds develop, they emerge
in pairs in perfect order. Tinges of pink
gradually appear, harmonising with
the pale silvery green.

I found this period in the cycle,
represented by this and the next painting,
to be particularly beautiful.

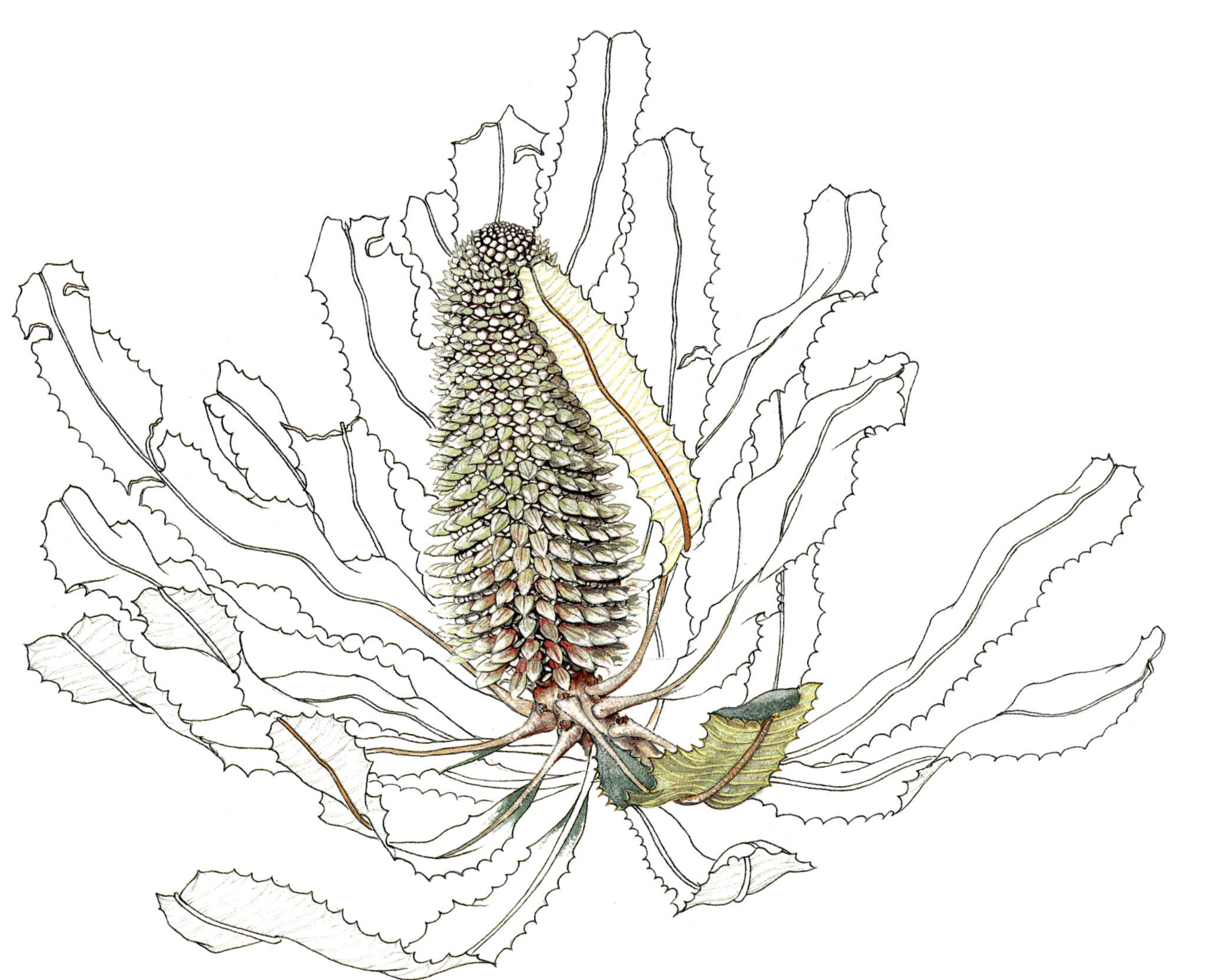

From the bottom to the top the buds
continue to grow, and to progressively
display more and more colour.

These stages were so captivating that
I found myself standing in front of the tree,
talking to it. I find it hard to articulate the
very intense feelings I had gazing at the
absolute perfection of the structure
of the budding spike.

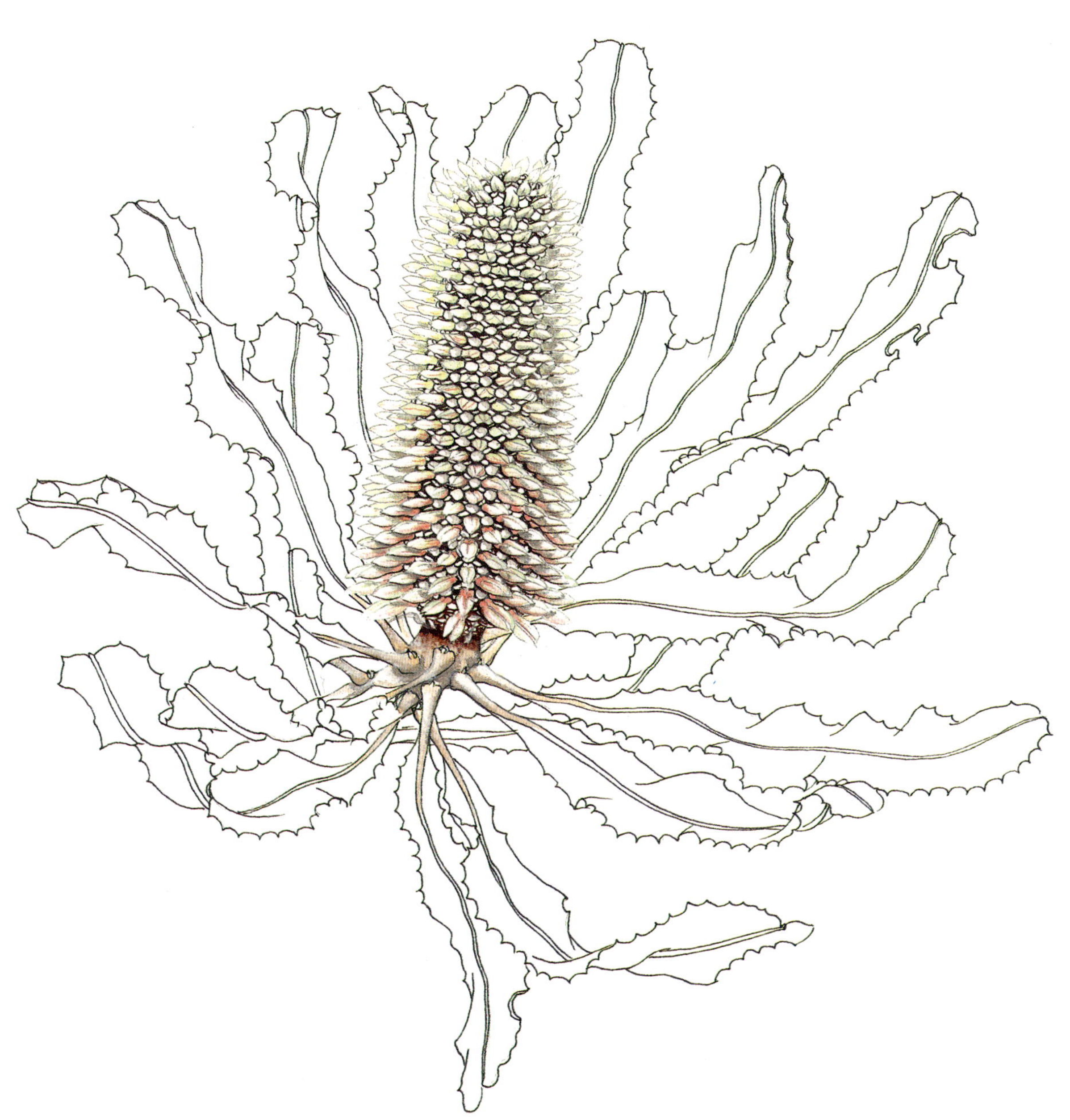

At this time I found the banksia's cycle
particularly intriguing and exciting. With
the further development of the budding
flowers, the orderly pattern on the spike
seemed to transform and become chaotic.

In all the trees I observed over a large area
(as far apart as two hundred kilometres),
this change to the pattern occurred at
precisely the same time – at any given
moment the transformation was in
exactly the same position, and at
the same rate on every spike.

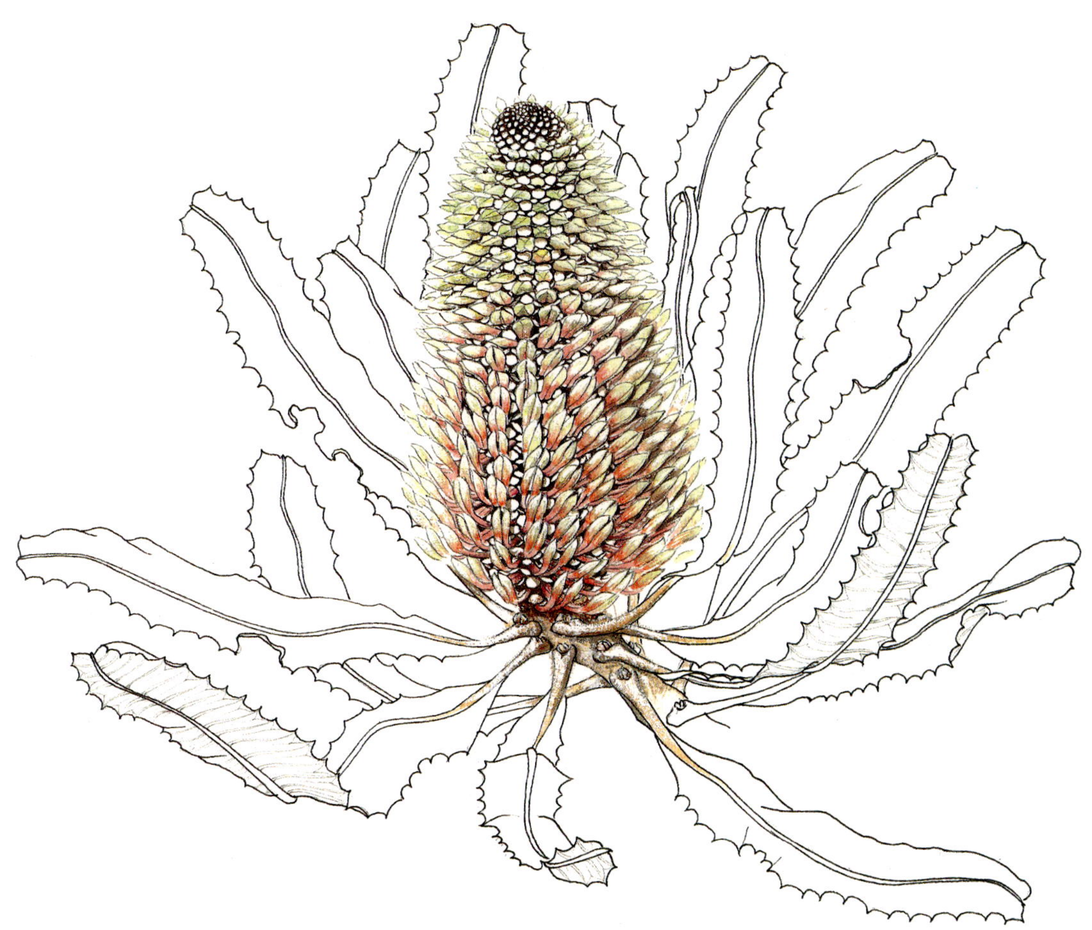

The pairs of flowers have now separated,
and whilst up close the arrangement
still seems relatively chaotic, viewed as
a whole the flowering spike becomes
tranquil, delicate and unified.

The leaves too created a special fascination
for me. I found that the elegance of the
patterns on the leaves required their own
attention. And the colour emerging from
the pencil and ink creates movement from
one picture into the next. The eye follows
the leaves, which leads inevitably to the
next stage, maintaining a continuity.

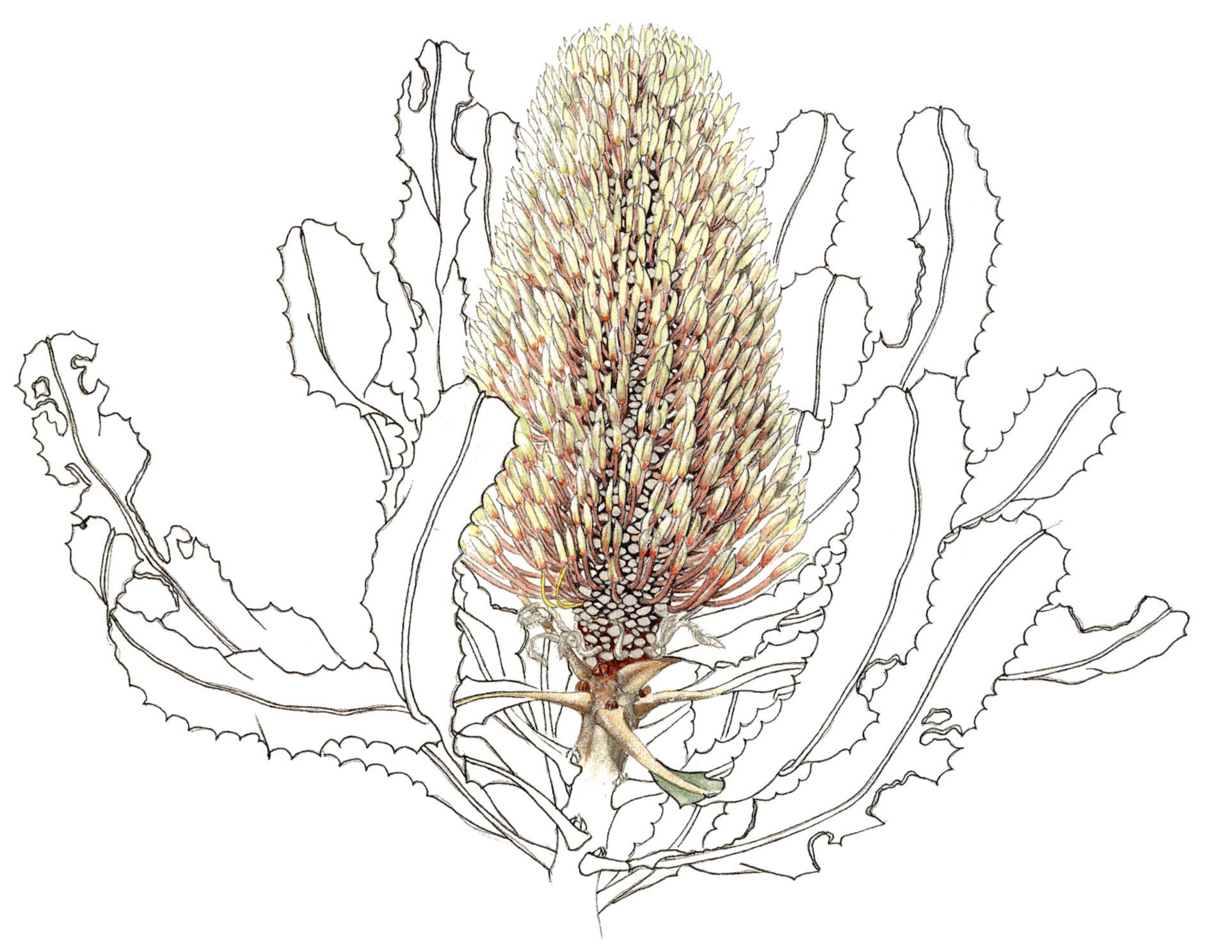

As the buds grow longer the dominant
silvery green colour gives way to a soft pink.

For me this is an image which is charged
with intense emotion – before it my body
relaxes and I become totally calm. I imagine
it could be a subject for deep meditation.

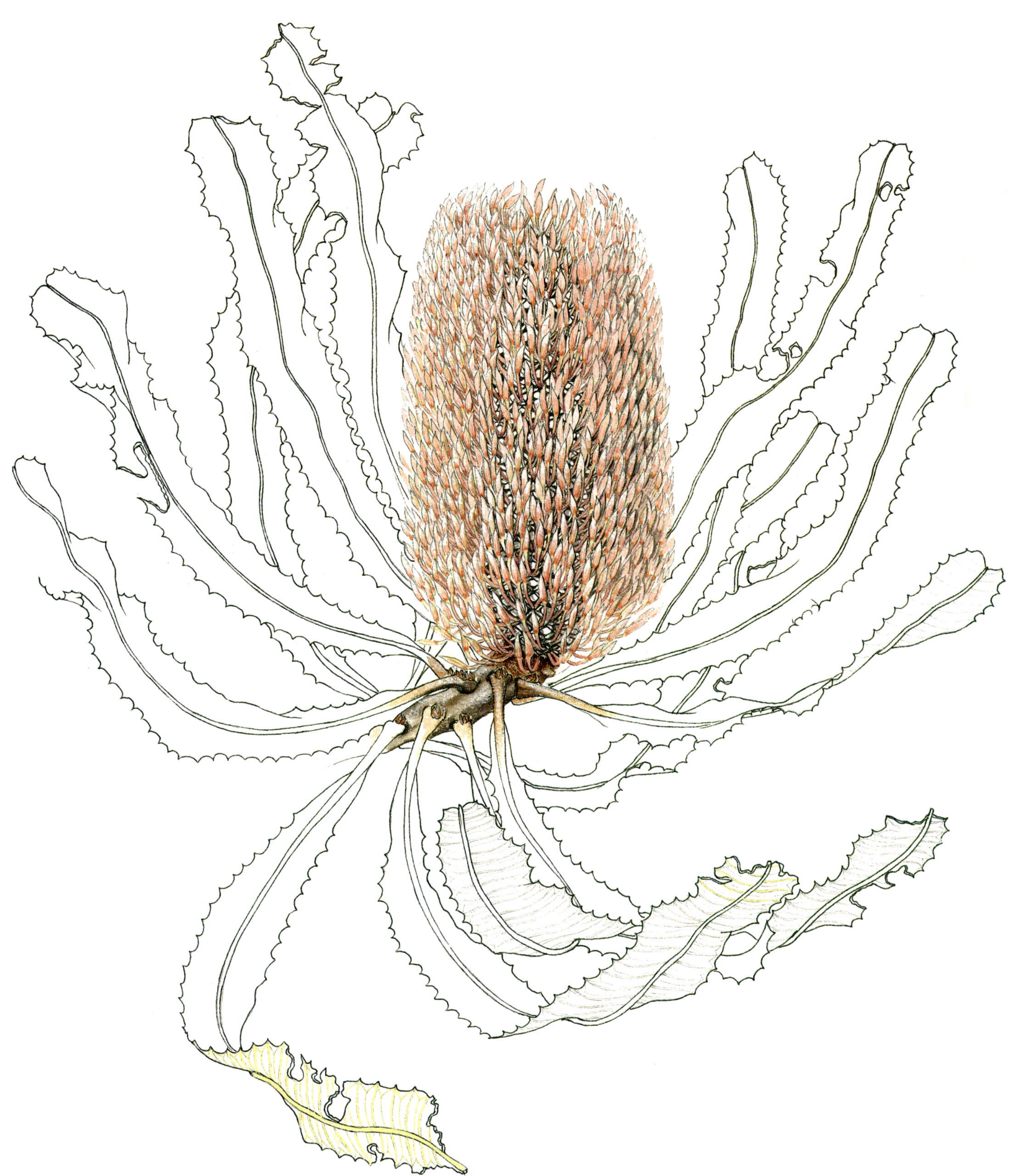

Amazingly the flowers now pair up again and form an ordered pattern once more. The soft, delicate pink hue has started to darken. The flowers also begin to develop their characteristic hook shape as a result of the enlargement of the pollen presenter.

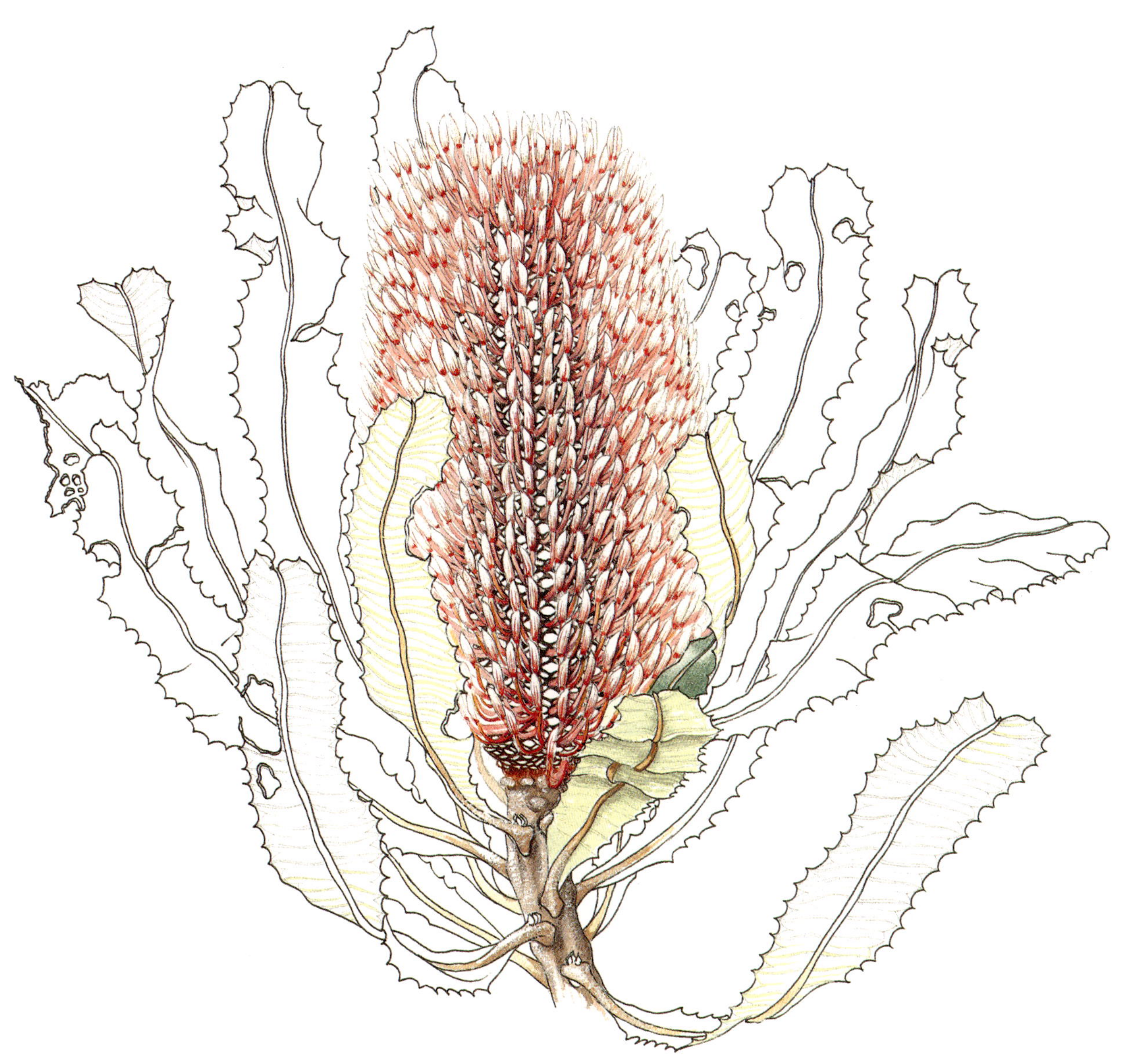

The flowers begin to bloom from the
bottom, again spiralling upwards, whilst
over the rest of the spike the unopened
flowers remain tightly ordered and are now
glowing rich alizarin crimson from within.

Tight, bright and looking to me like
beacons, the flowers are already
filling with ants and bees.

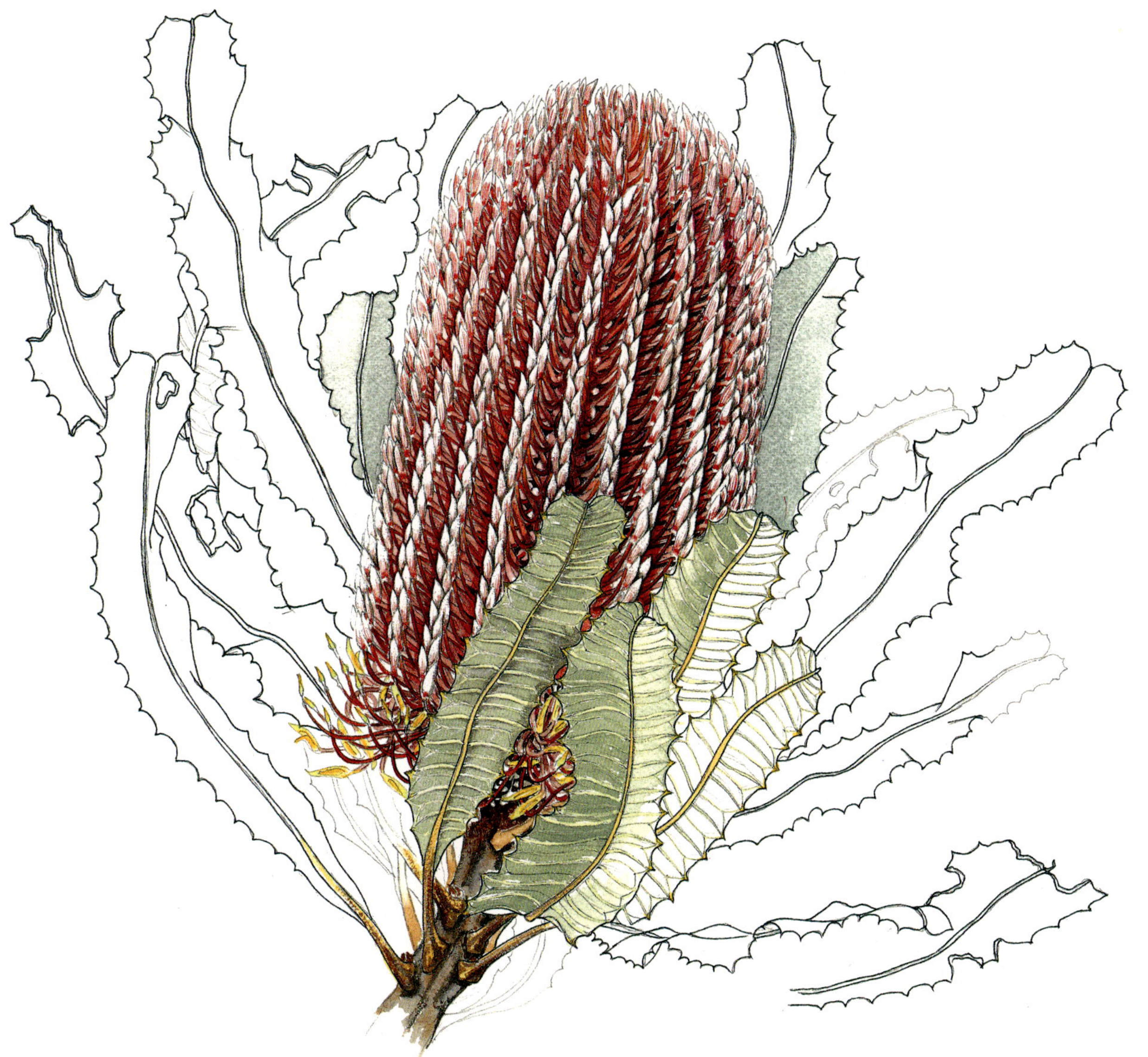

This was to be the grand statement –
the *Banksia menziesii* in all its glory.
However, in looking through all the
paintings I came to realise how each step
fulfils its part equally in the whole.

I am now not at all sure that I am as
attracted to this well-known image of the
species as much as I am to some of the
others in this series. A number of other
images stimulate a much stronger
emotional response in me, and
therefore have more meaning.

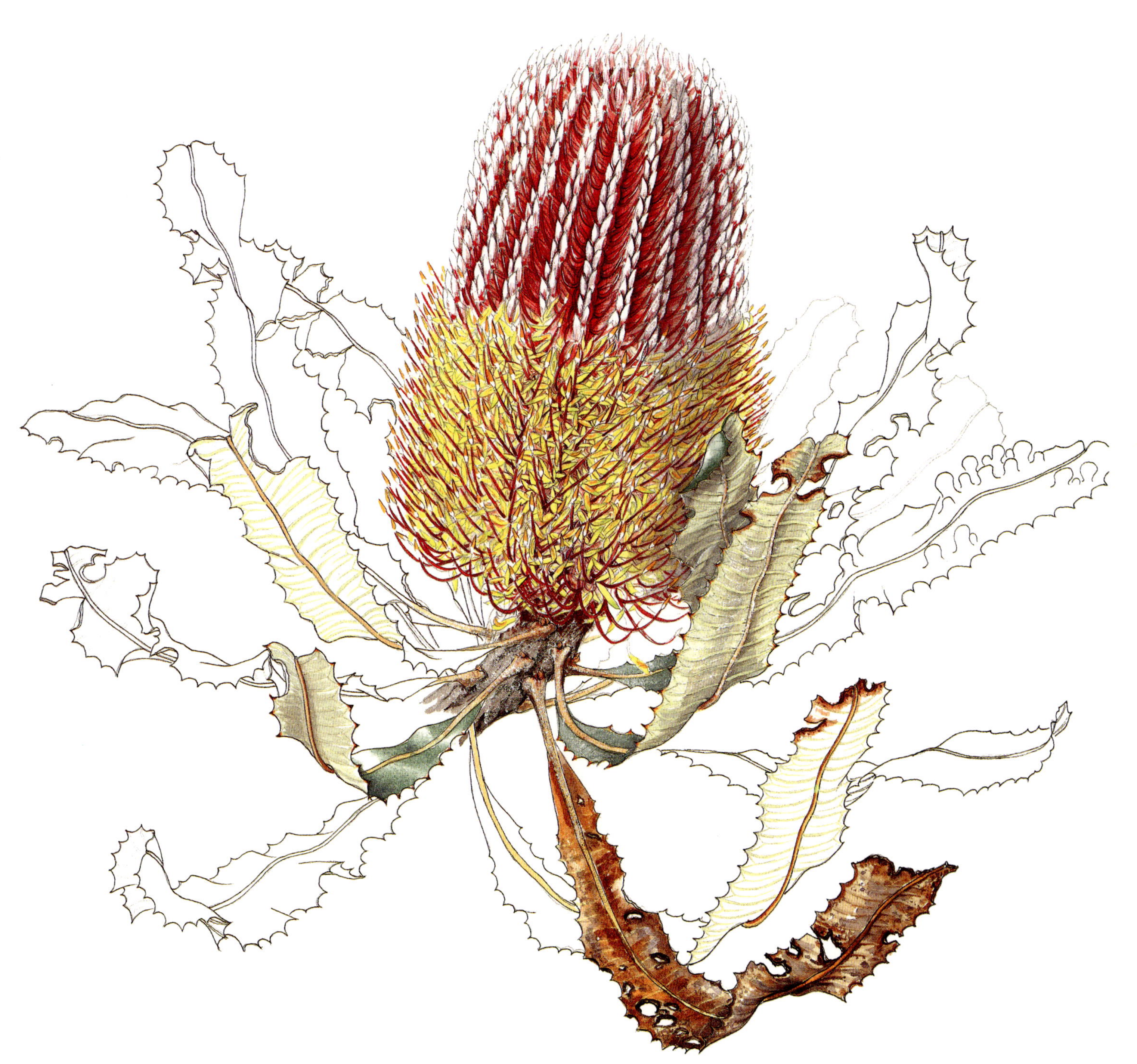

The flowers continue to open all the way
up the spike. In full bloom there are
thousands of flowers (five to six thousand
per spike are possible) awaiting pollination
– potentially a myriad of new banksias.

A time of transition, as well as a point
of arrival. I was now waiting for the
steps which were to follow.

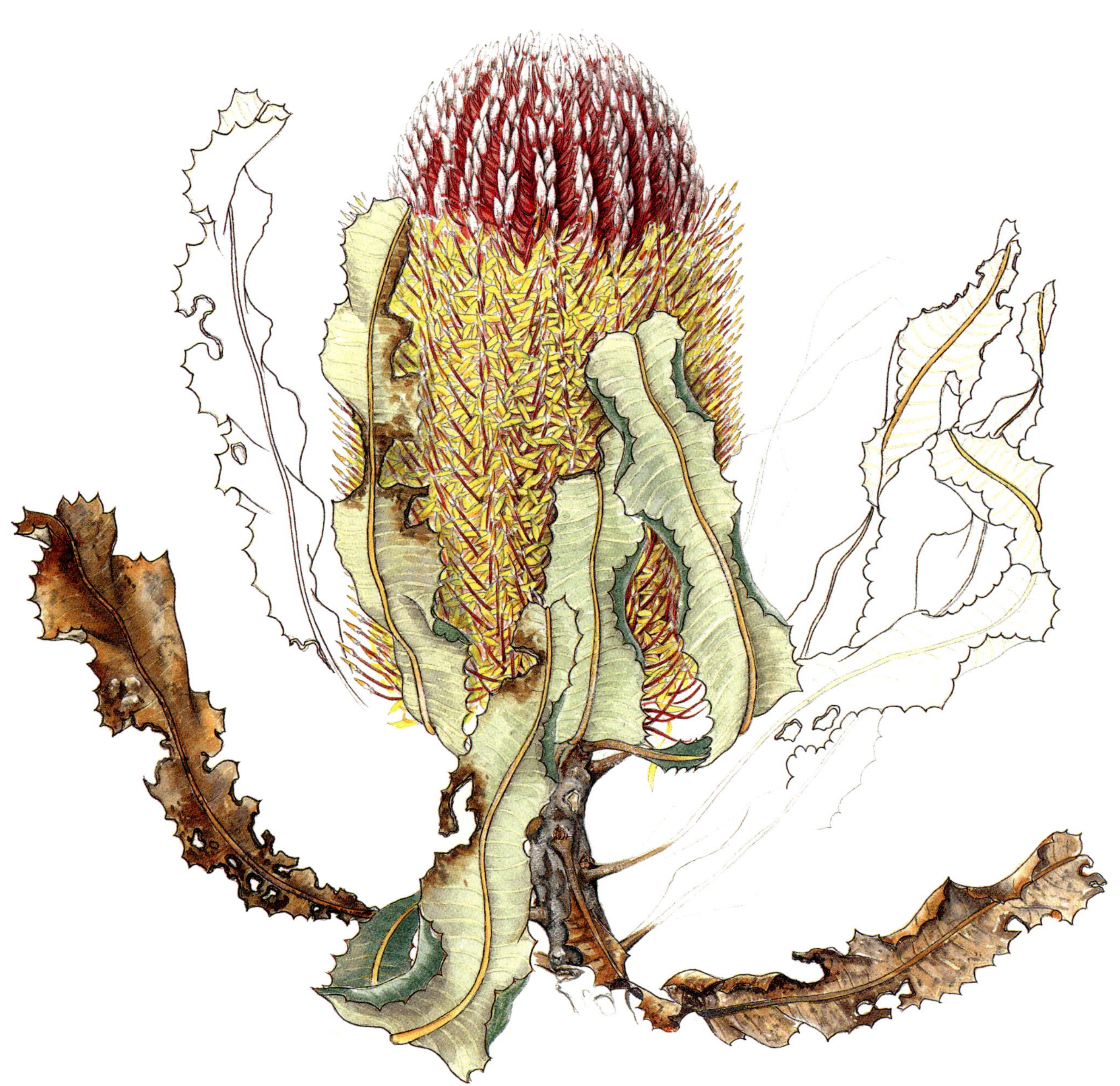

Just a lot of dead and dying flowers?
Each of the pollinated flowers is the source
of incredible potential – new seeds, which
in their turn can become new life.

Looking at this stage of the banksia's
cycle, for me there is a sense of growing
suspense, of anticipation.

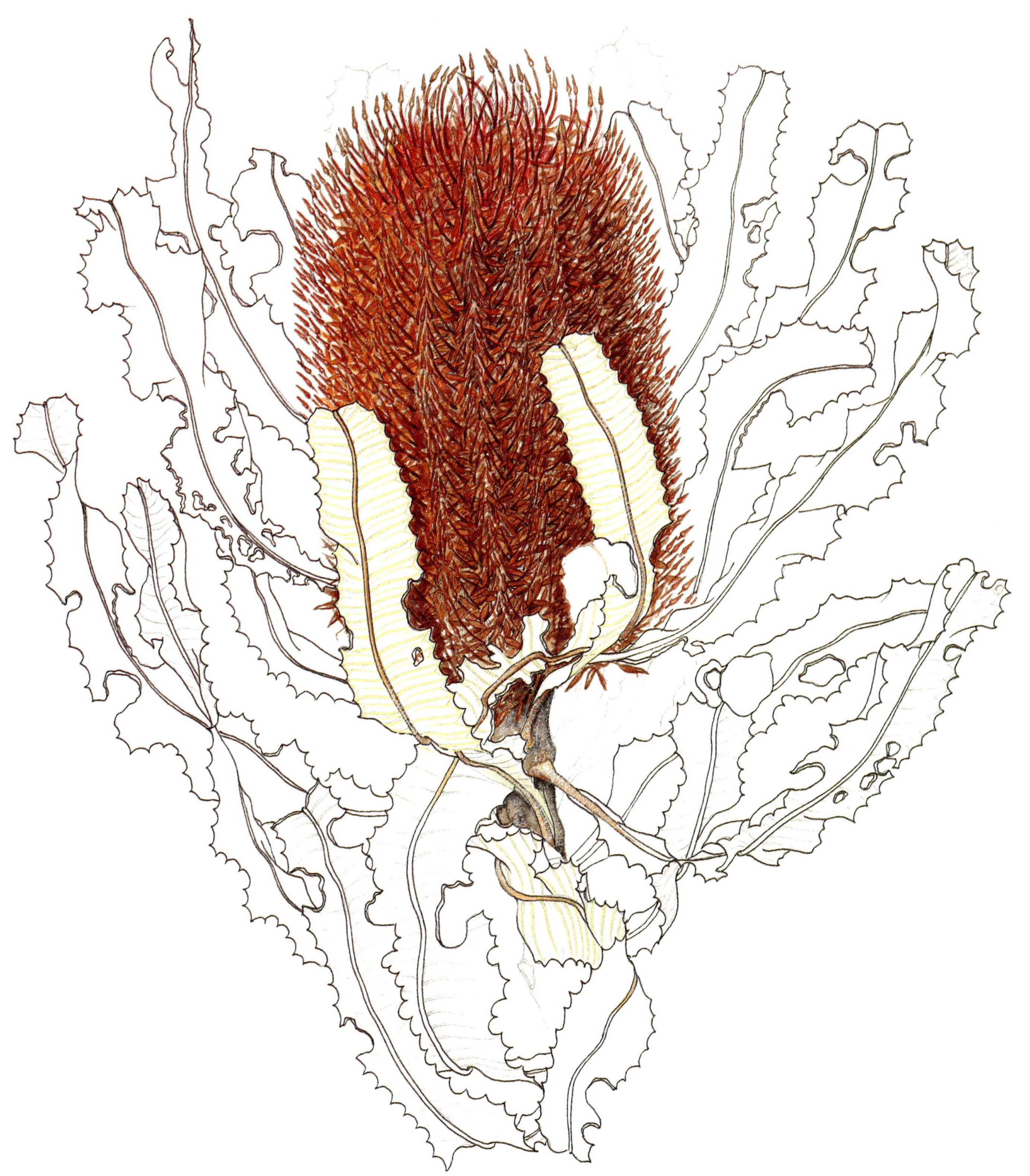

The extraordinary tangle of dead flowers
fall away to reveal again the perfectly
ordered patterns beneath. Rather than the
soft and delicate spike at the beginning of
its development, it is now a hard intricately
patterned cone, covered in a velvety
down of browns and greys.

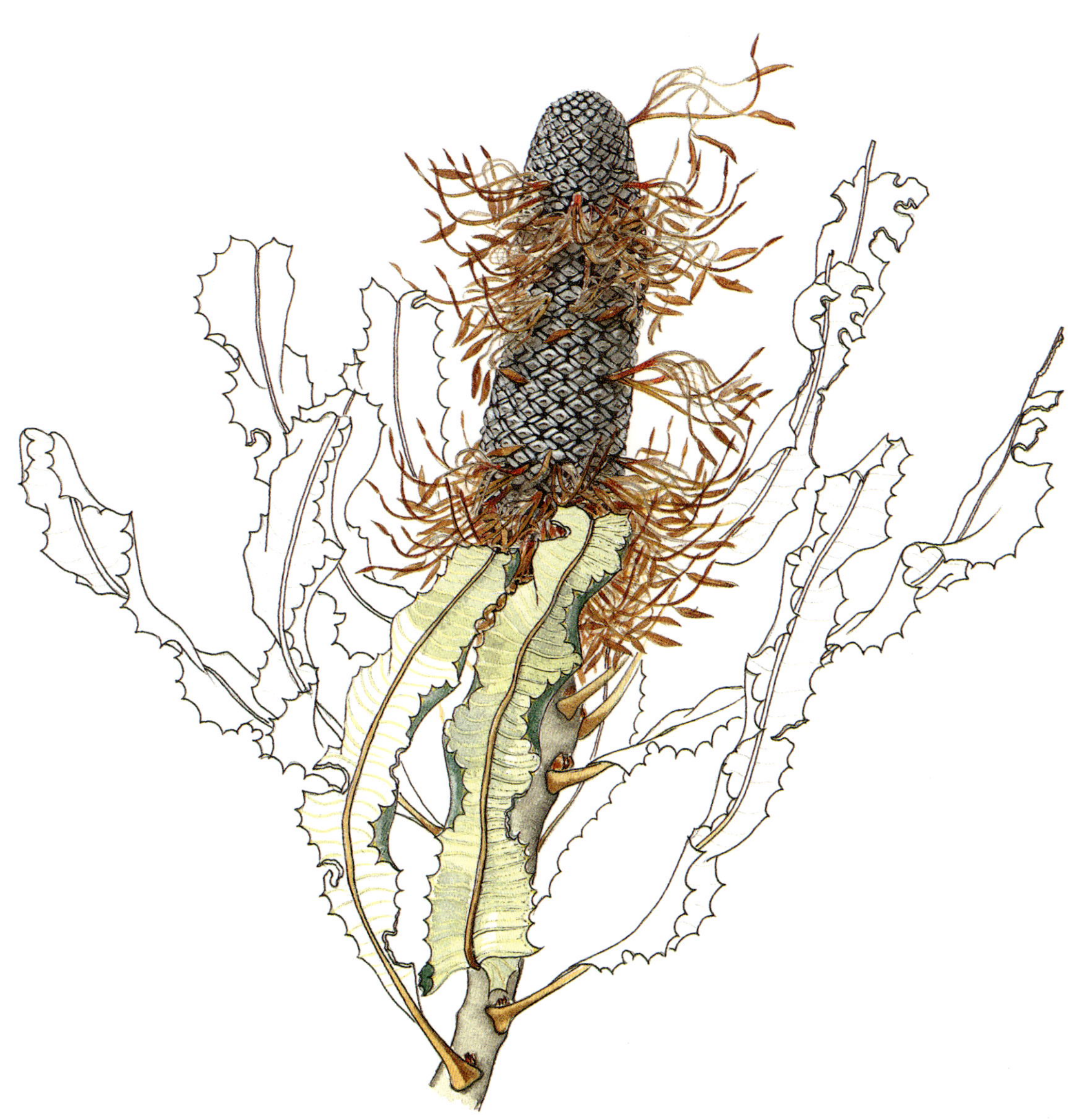

Some of the patterns on the cone are displaced as the beginnings of the seed-bearing follicles appear. At this point only tantalisingly small tips can be seen, appearing as if by magic. Of all those thousands of flowers just this small number were pollinated.

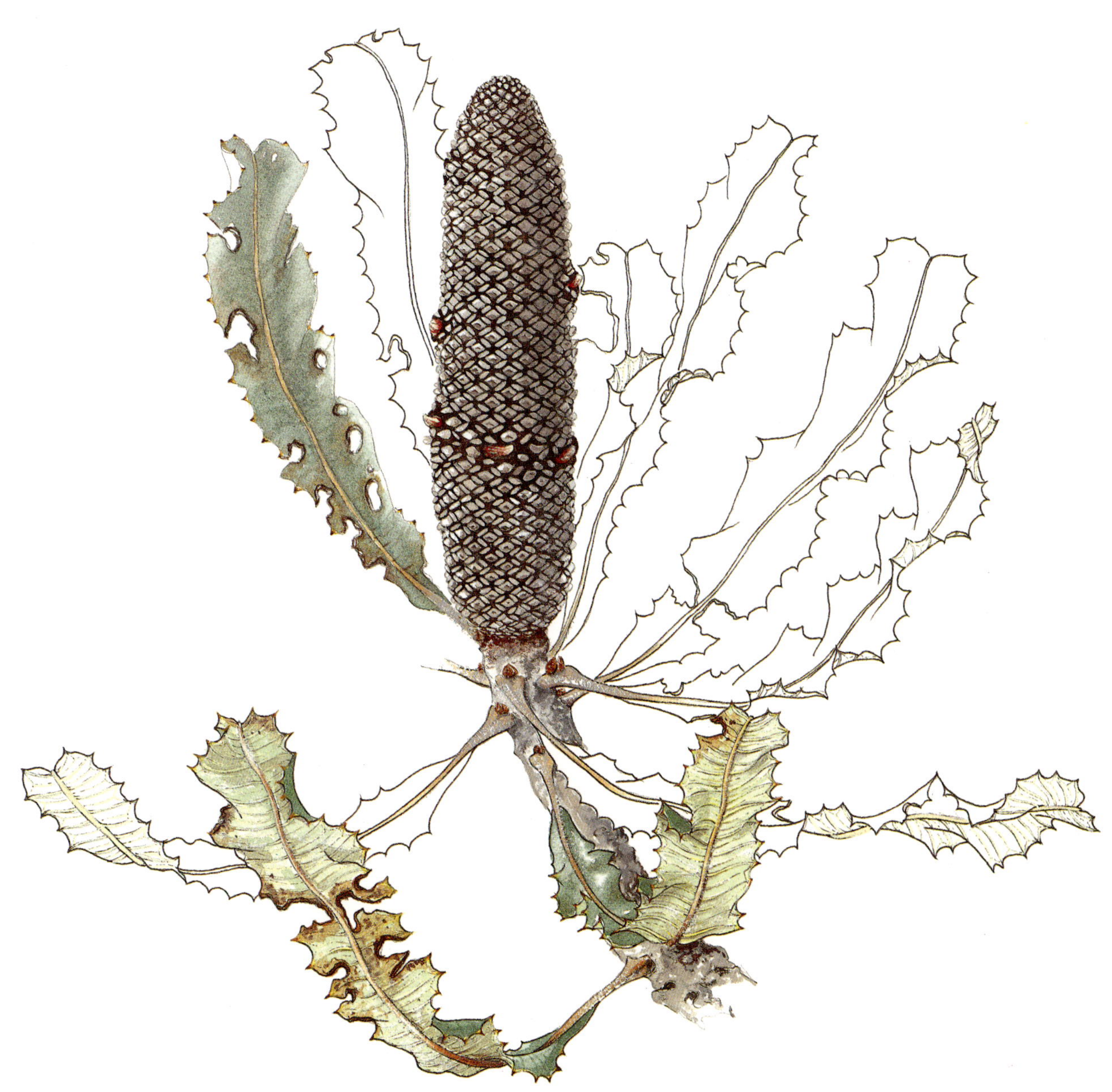

As the follicles grow and push out new
colours emerge, and the patterns around
them on the surface of the spike transform
to new alignments to accommodate the
swelling follicles. The follicles, in fact,
appear to me as if they are already new
plants, for they are sap green and tender as
a new plant. Another cycle within a cycle.

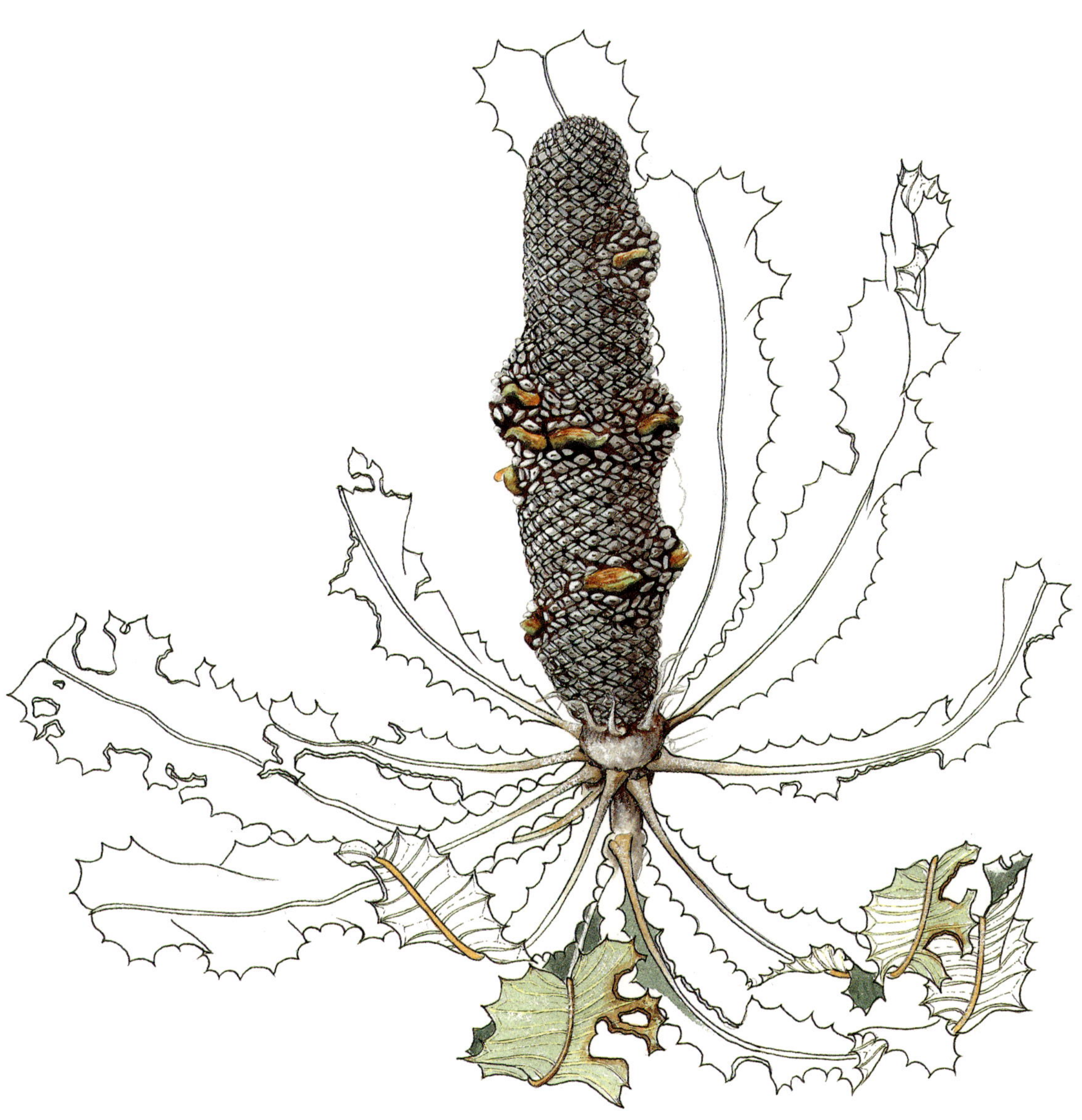

As the follicles become further developed
the shape of the cone becomes distorted.

And even as the seeds within are enlarging,
from the base of the fruiting cone, a new
shoot has appeared, giving the plant two
chances for renewal – seed and bud.

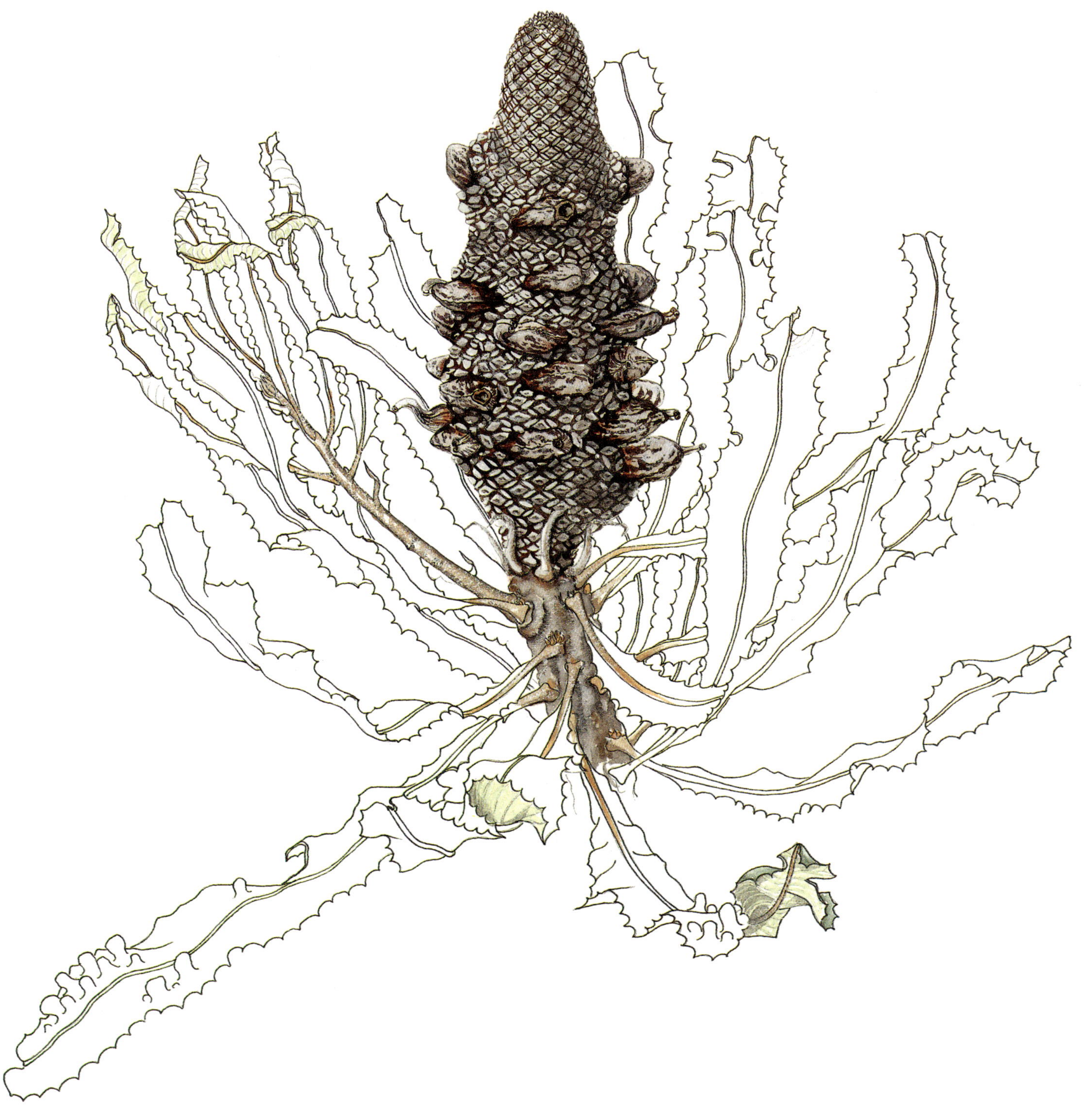

At this stage I find I am holding my breath…
waiting. The swollen follicles are full of
tension, they appear about to burst.
I am also attracted by the colours and
tones of the fluffy coating of the follicles.

Meanwhile, the new stem is pushing out
and developing the next year's spike bud.

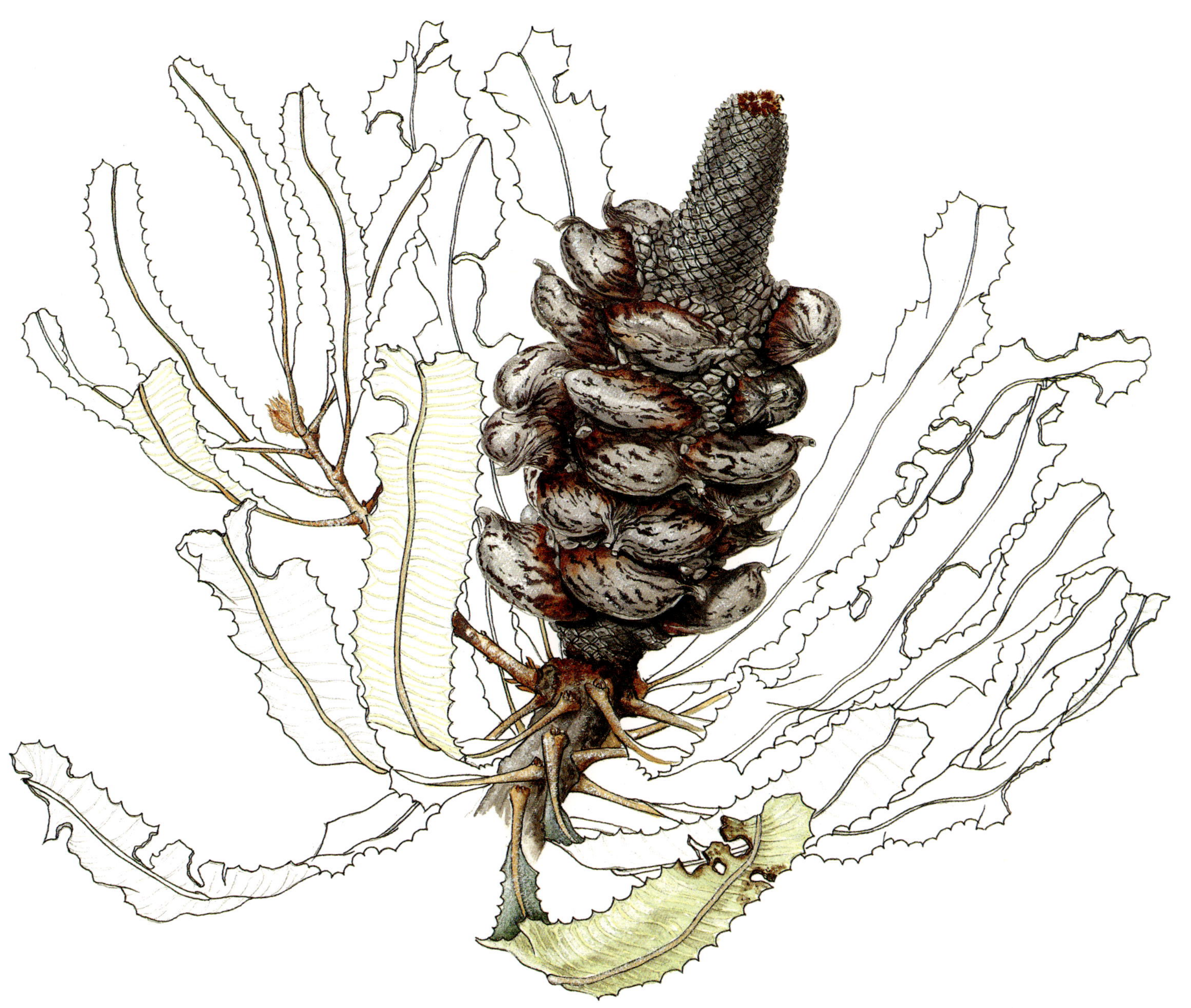

The follicles have split open, the seeds
flown on the winds, the work of the
mother plant completed. The empty cone,
with its rich earth colours sits in spent
glory among next year's potential flowers.

As a subject for drawing I find this stage
most captivating, with its echoes of
May Gibb's Banksia men.

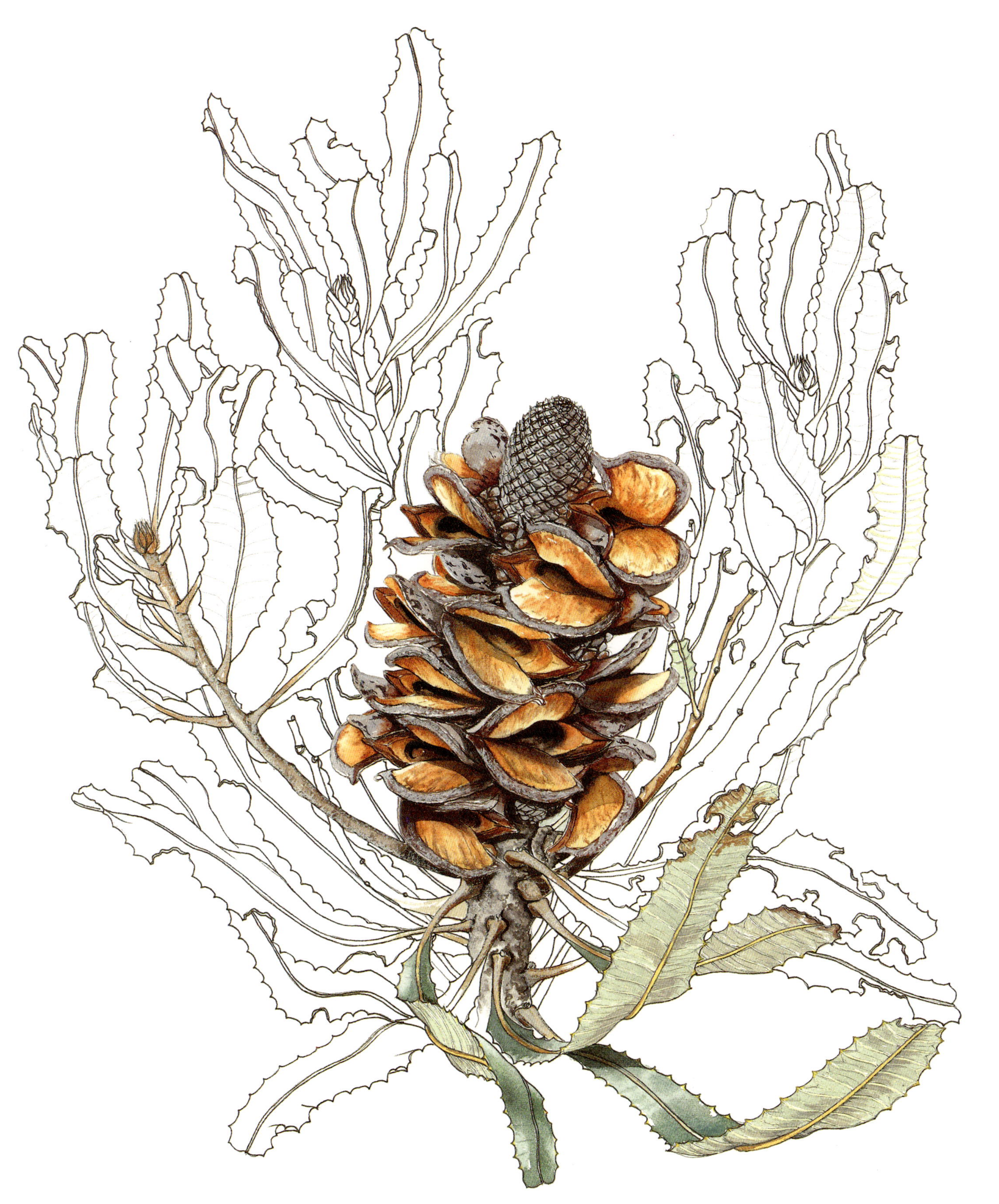

The beautiful, earth-coloured seeds are
thrown from the mother plant, are carried
on the wind, and have fallen to the earth.
So aged-looking, yet the essence of new life.

For the *Banksia menziesii* the cycle from
seedling to the first flowering is six to ten
years. Thereafter, under normal conditions
each tree flowers annually from
February to August.

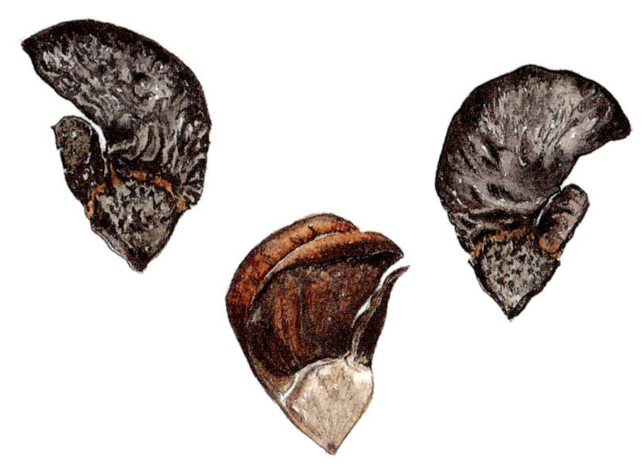

The seed combines with the elements to
send forth roots and dicotyledonous leaves
– new life begins again. Seeing this renewal
year in year out, the mediaeval concept
of cyclical time has more meaning for
me than the more recent construct
of continuous time.

A year after completing the paintings
I find that my excitement at seeing the
whole cycle again has not abated, it remains
as overwhelming as ever. As I look at them
I know the trees are again in the process of
flowering and the desire to take off to
commune with them returns. The bush for
me has always been a source of renewal.

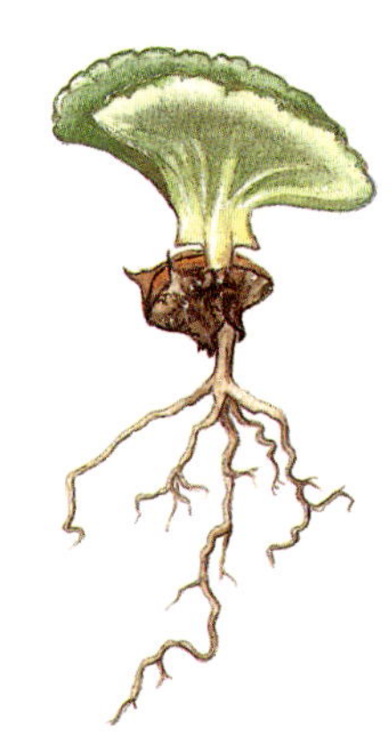

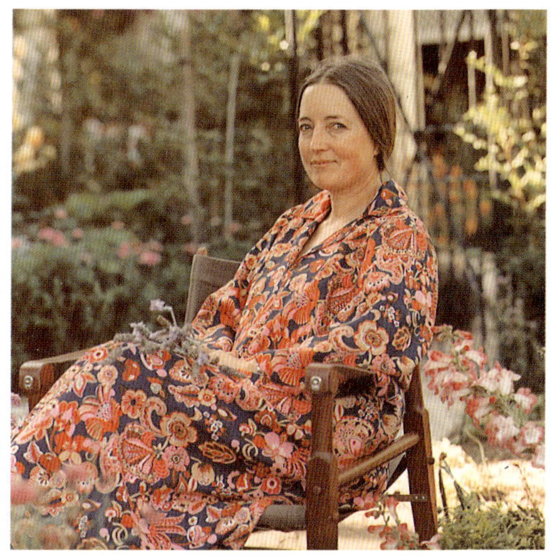

Philippa Nikulinsky was born in Kalgoorlie, Western Australia, in 1942. She trained as an art teacher and has taught in a number of secondary schools and tertiary institutions. She now works full-time as a natural history artist and illustrator, specialising in Australian native flora.

Philippa Nikulinsky's work has appeared in numerous group and solo exhibitions throughout Australia, and she has received many major public and corporate commissions in Australia and overseas. Her work has also been included in a wide variety of books and journals. She has previously published two books – *Western Australian Wildflowers in Watercolour* and *Flowering Plants of the Eastern Goldfields of Western Australia* – a limited edition of prints and a number of posters. Two books, *Banksia menziesii* and *The Australian Wildflower Diary 1993* are being published by Fremantle Arts Centre Press in 1992.

Photograph by Pat Barblett